I0846327

No part of this publication may be reproduced, distributed, or transmitted in any form or by any means, including photocopying, recording, or other electronic or mechanical methods nor may any other exclusive right be exercised, excluding any use permitted under the Copyright Act 1968 as amended, without the prior written permission of the author and illustrator.

The story, all names, characters, and incidents portrayed in this production are fictitious. No identification with actual persons, places, buildings, and products is intended or should be inferred.

Copyright © 2023, Charlotte Wynter
Copyright © 2023, Sarah Wan

All rights reserved.

Paperback ISBN 9798861840569

Illustrations by Sarah Wan were created digitally.

A MATHEMATICAL JOURNEY

Perri's Adventure Home

Written by Charlotte Wynter

Illustrated by Sarah Wan

Across the native grassland on the hard red-brown soil, Perri the plains-wanderer dashed.

She had been wandering across the great Outback and was
on her way home with a souvenir.

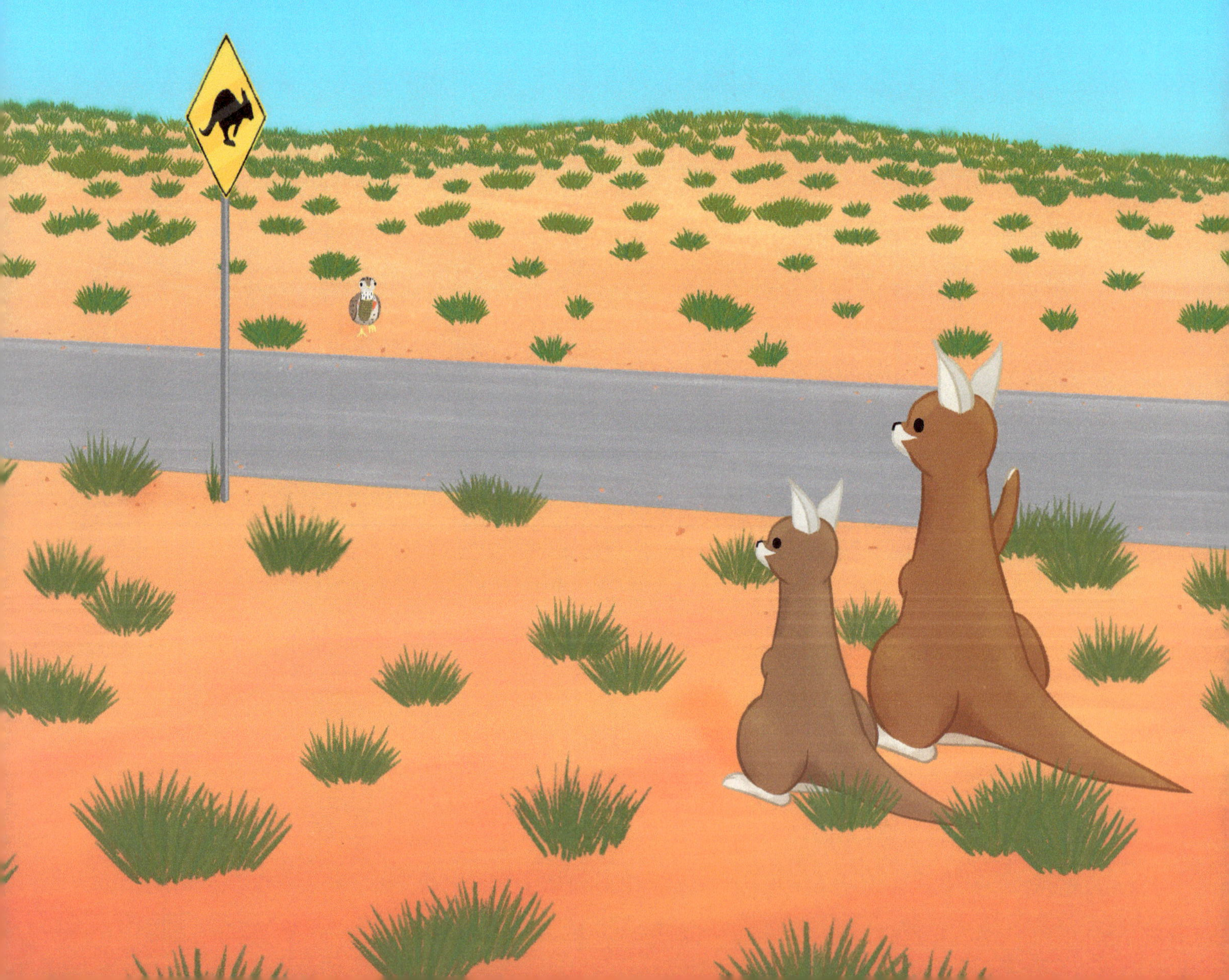

"Hi Perri," Jumpy waved. "How was your trip?"
"It was great, I even brought back some ochre," she replied with a smile.
"How many did you manage to find?"
"I'm not sure how many pieces I collected."

"Let's have a look," Jumpy said and hopped over.
As Perri walked over, she tripped. Jumpy watched the stones roll out
in a straight line across the grey concrete path.

"This is going to take a while," sighed Perri, feeling deflated.

"How about marking them in groups of ten?" Bouncy suggested.

"Okay, let's try!"

1...2...3...4...5...6...7...8...9...10

Perri went along and counted to ten before placing a stick as a marker.

After she marked the last group of ten, she called, "Ready!"

Jumpy jumped to each marker. "10, 20, 30, ..., 110, 120 pieces!" they cheered. "Amazing!"

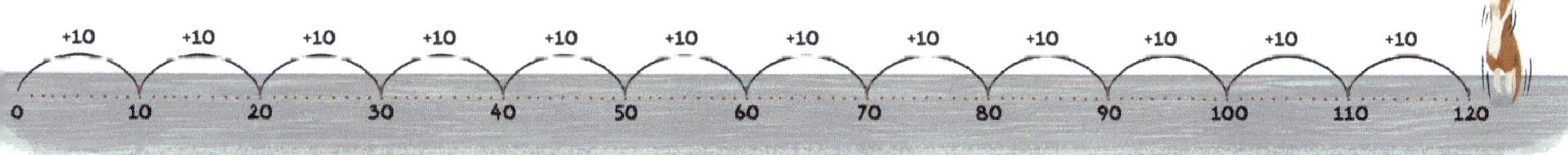

"Thanks Jumpy. The both of you can take forty-two," Perri smiled.

"Thank you," Bouncy said. "How many will you ahve left?"

"Hmm, I'm not sure," Perri began to worry.

"I know what to do!" Bouncy bounced with excitement. He quickly hopped to the end of the line.

They watched as Bouncy started from the end of the line and did

four **large jumps** then two *little jumps.*

"You gave us forty-two pieces. Now you have seventy-eight pieces of ochre left," Bouncy told Perri.

Jumpy and Bouncy jumped and gave each other a high-five.

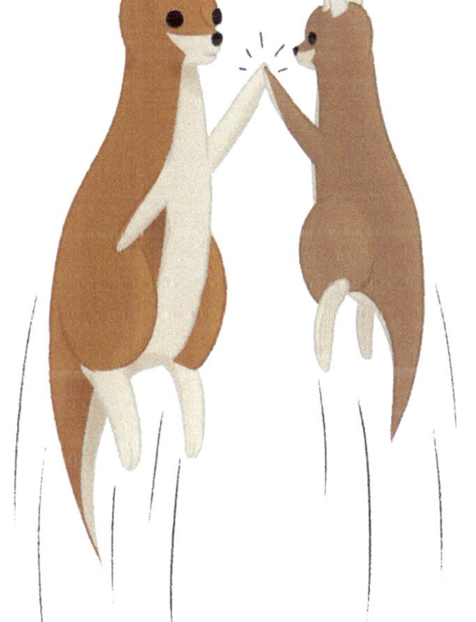

Perri thanked them again and continued on her route.

Soon, she came close to a familiar tree, which was also the home of her good friends Wattle and Ash.

The two little budgies were flying with beautiful, pink and yellow gumnut blossoms in their small, curved beaks.

"How many did you find?" Wattle asked.

"Thirty-three. How about you?" Ash replied.

"Twenty-five."

"So how many do we have all together?"

They immediately started splitting their piles of gumnuts and then joined them together into one large sum.

They put the last of their gumnuts into the pile. "Fifty-eight!" they cheered.

"Hello," Perri called up to them as she approached. " I brought back chunks of ochre from my trip across the Outback. Would you like some?"

Wattle and Ash looked at each other then back to Perri and replied, "We'd like thirty-seven please. Enough to share with the whole flock."

Perri began to take out the ochre but a puzzled look crossed her face.

"I have seventy-eight pieces of ochre but after I give you thirty-seven, how many will I have left?"

"We can help!" exclaimed Wattle and Ash.

"Seventy minus thirty is forty," began Wattle as he settled on the ground.
"And eight minus seven equals one," followed Ash.

Then, together they chirped out gleefully. "That means you will have forty-one pieces of ochre left!"
Perri tilted her head, still unsure.

"That's okay. We'll show you."

They flew up, swooping about like they were in a dance.

Their claws picked up the ochre...

...to drop them on green eucalyptus leaves.

Followed by the use of their tiny beaks to...

pull leaves

and
push ochre.

78   -   37
70 - 30   +   8 - 7
40   +   1
I see.

Perri clapped her wings, "I see, that makes sense. Thank you for showing me. That's a great strategy."

Wattle and Ash chirruped.They said farewell to Perri and took the ochre to the skies to share with their family.

So off Perri went, through the bush.

Under the midday sun, Perri saw a lizard basking in the open plains.

It was dark brown with scales like a pine cone
and it had a short, stumpy, triangular head.

"Hi, Jana!" Perri greeted.

"G'day Perri," a voice answered from her right.

Perri jumped back in surprise and looked up.

She saw Jana smiling at her from the other end of her tail.

"You scared me!" Perri laughed.

"Sorry about that," she apologised.
"It's good to see you. How was the trip?"

Perri told her about the amazing views of bright, colourful, layered rock. It was like looking at waterfalls of gold, orange, brown and crimson shining softly in the sun.

"As promised, I brought back some ochre for you."

"Thank you for giving some to me, Perri."

"I have forty-one pieces and I need to make sure I bring eighteen pieces back to the nest. The difference is...?" Perri paused to think.

"Eighteen plus two makes twenty. Add two tens makes forty. Then, plus one makes forty-one."

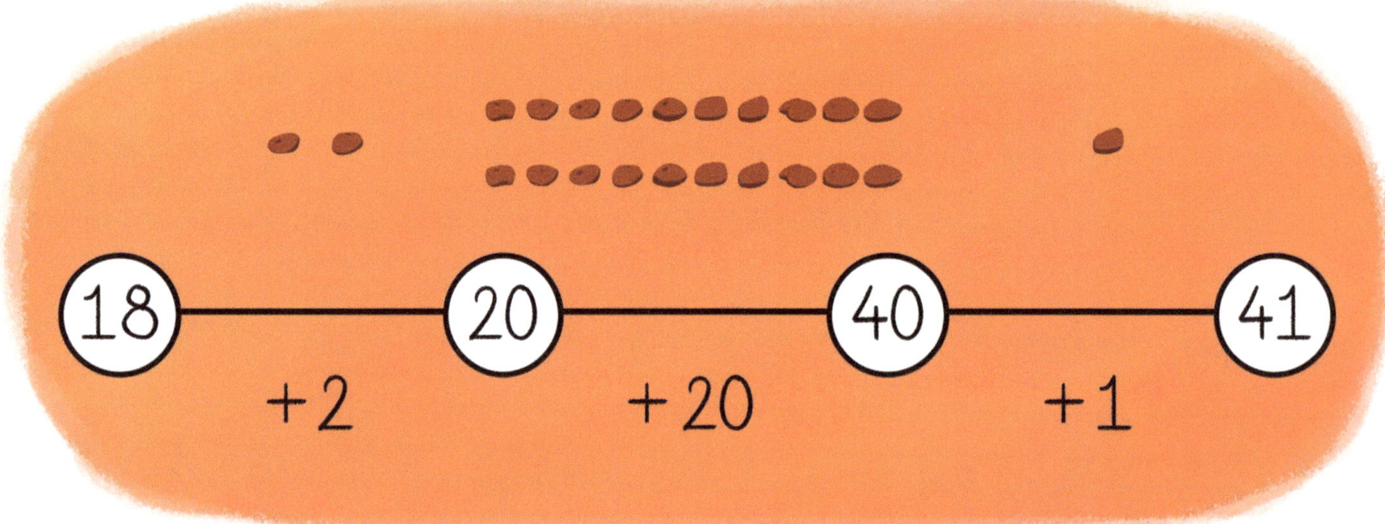

"So, two plus ten, plus ten, plus one."

$$41 = 18 + \boxed{?}$$
$$= 18 + (2 + 20 + 1)$$
$$2 + 20 + 1 = 23$$

"That's the same as two plus twenty, plus one, which equals twenty-three."

Jana thankfully swept
the stones into a hole
with her bobtail.

Perri happily took the remaining ochre pieces
and waved goodbye to her friend.

Finally, Perri returned home to a clutch of excited chicks.

THE END

www.ingramcontent.com/pod-product-compliance
Lightning Source LLC
Chambersburg PA
CBHW040209240726
48664CB00002B/892